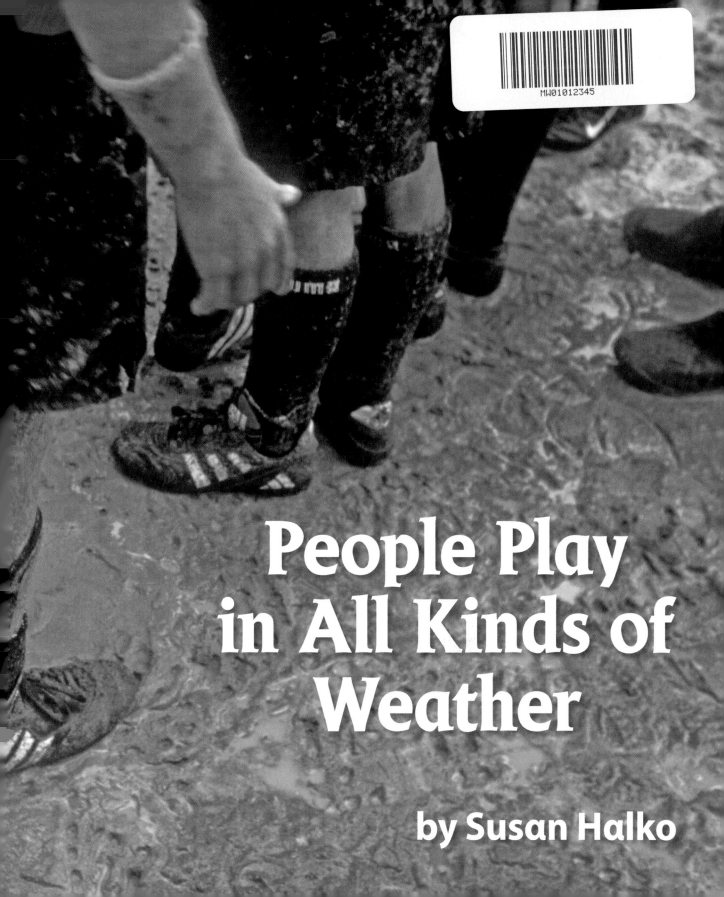

People Play in All Kinds of Weather

by Susan Halko

People fly kites in **windy** weather.

3

People swim in **sunny** weather.

People splash in **rainy** weather.

People jump in **cloudy** weather.

People sled in snowy weather.

People Play in All Kinds of Weather

windy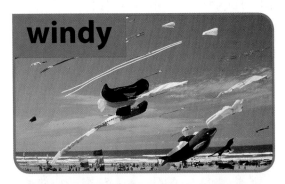

cloudy

sunny

snowy

rainy